THIS BOOK BELONGS TO

EMAIL:

ADDRESS:

CONTACT:

PHONE:

START DATE	END DATE

MO TU WE TH FR SA SU
☐   ☐   ☐   ☐   ☐   ☐   ☐

DATE: _____ / _____ / _____

PROJECT:

FOREMAN:

WEATHER

F° _____   C° _____   _____ AM   _____ PM

HOURS DUE TO BAD WEATHER	ISSUED AND DELAYS

NOTE: _______________________________________

COMPLETION DATE	DAYS AHEAD OF SCHEDULE	DAYS BEHIND SCHEDULE

SAFETY AND INCIDENTS

SAFETY ISSUES THAT NEED TO BE ADDRESSED	ACCIDENTS / INCIDENTS / STEPS NEEDED TO RESOLVE

SUMMARY OF THE WORK DONE TODAY

IMPORTANT NOTES

NAME	SIGNATURE

TODAY LABOR

INITIALS	TRADE	START	FINISH	PAID HOURS	OVERTIME	COMPANY
☐ EMPLOYEE ☐ CONTRUCTOR		AM	PM			
☐ EMPLOYEE ☐ CONTRUCTOR		AM	PM			
☐ EMPLOYEE ☐ CONTRUCTOR		AM	PM			
☐ EMPLOYEE ☐ CONTRUCTOR		AM	PM			
☐ EMPLOYEE ☐ CONTRUCTOR		AM	PM			
☐ EMPLOYEE ☐ CONTRUCTOR		AM	PM			
☐ EMPLOYEE ☐ CONTRUCTOR		AM	PM			
☐ EMPLOYEE ☐ CONTRUCTOR		AM	PM			

EQUIPMENT ON SITE	NO. OF UNITE	WORKING YES / NO

HIRED EQUIPMENT	NO. OF UNITE	EQUIPMENT RENTED	FROM	RATE

NAME: _______________________   SIGNATURE: _______________________

MO   TU   WE   TH   FR   SA   SU
☐   ☐   ☐   ☐   ☐   ☐   ☐

DATE: /   /

PROJECT:

FOREMAN:

WEATHER

F°   C°   AM   PM

HOURS DUE TO BAD WEATHER	ISSUED AND DELAYS

NOTE:

COMPLETION DATE	DAYS AHEAD OF SCHEDULE	DAYS BEHIND SCHEDULE

SAFETY AND INCIDENTS

SAFETY ISSUES THAT NEED TO BE ADDRESSED	ACCIDENTS / INCIDENTS / STEPS NEEDED TO RESOLVE

SUMMARY OF THE WORK DONE TODAY

IMPORTANT NOTES

NAME	SIGNATURE

TODAY LABOR

INITIALS	TRADE	START	FINISH	PAID HOURS	OVERTIME	COMPANY
☐ EMPLOYEE ☐ CONTRUCTOR		AM	PM			
☐ EMPLOYEE ☐ CONTRUCTOR		AM	PM			
☐ EMPLOYEE ☐ CONTRUCTOR		AM	PM			
☐ EMPLOYEE ☐ CONTRUCTOR		AM	PM			
☐ EMPLOYEE ☐ CONTRUCTOR		AM	PM			
☐ EMPLOYEE ☐ CONTRUCTOR		AM	PM			
☐ EMPLOYEE ☐ CONTRUCTOR		AM	PM			
☐ EMPLOYEE ☐ CONTRUCTOR		AM	PM			

EQUIPMENT ON SITE	NO. OF UNITE	WORKING YES / NO

HIRED EQUIPMENT	NO. OF UNITE	EQUIPMENT RENTED	FROM	RATE

NAME: _______________________   SIGNATURE: _______________________

MO TU WE TH FR SA SU
☐   ☐   ☐   ☐   ☐   ☐   ☐

DATE: _____ / _____ / _____

PROJECT:

FOREMAN:

WEATHER

F° _____   C° _____   _____ AM   _____ PM

| HOURS DUE TO BAD WEATHER | ISSUED AND DELAYS |

NOTE: _______________________

| COMPLETION DATE | DAYS AHEAD OF SCHEDULE | DAYS BEHIND SCHEDULE |

SAFETY AND INCIDENTS

| SAFETY ISSUES THAT NEED TO BE ADDRESSED | ACCIDENTS / INCIDENTS / STEPS NEEDED TO RESOLVE |

SUMMARY OF THE WORK DONE TODAY

IMPORTANT NOTES

| NAME | SIGNATURE |

TODAY LABOR						
INITIALS	TRADE	START	FINISH	PAID HOURS	OVERTIME	COMPANY
☐ EMPLOYEE ☐ CONTRUCTOR		AM	PM			
☐ EMPLOYEE ☐ CONTRUCTOR		AM	PM			
☐ EMPLOYEE ☐ CONTRUCTOR		AM	PM			
☐ EMPLOYEE ☐ CONTRUCTOR		AM	PM			
☐ EMPLOYEE ☐ CONTRUCTOR		AM	PM			
☐ EMPLOYEE ☐ CONTRUCTOR		AM	PM			
☐ EMPLOYEE ☐ CONTRUCTOR		AM	PM			
☐ EMPLOYEE ☐ CONTRUCTOR		AM	PM			

EQUIPMENT ON SITE	NO. OF UNITE	WORKING YES / NO

HIRED EQUIPMENT	NO. OF UNITE	EQUIPMENT RENTED	FROM	RATE

NAME: _______________________   SIGNATURE: _______________________

MO   TU   WE   TH   FR   SA   SU
☐   ☐   ☐   ☐   ☐   ☐   ☐

DATE: /   /

PROJECT:

FOREMAN:

WEATHER

F°_____   C°_____   AM_____   PM_____

| HOURS DUE TO BAD WEATHER | ISSUED AND DELAYS |

NOTE: ____

COMPLETION DATE	DAYS AHEAD OF SCHEDULE	DAYS BEHIND SCHEDULE

SAFETY AND INCIDENTS

SAFETY ISSUES THAT NEED TO BE ADDRESSED	ACCIDENTS / INCIDENTS / STEPS NEEDED TO RESOLVE

SUMMARY OF THE WORK DONE TODAY

IMPORTANT NOTES

NAME	SIGNATURE

TODAY LABOR

INITIALS	TRADE	START	FINISH	PAID HOURS	OVERTIME	COMPANY
☐ EMPLOYEE ☐ CONTRUCTOR		AM	PM			
☐ EMPLOYEE ☐ CONTRUCTOR		AM	PM			
☐ EMPLOYEE ☐ CONTRUCTOR		AM	PM			
☐ EMPLOYEE ☐ CONTRUCTOR		AM	PM			
☐ EMPLOYEE ☐ CONTRUCTOR		AM	PM			
☐ EMPLOYEE ☐ CONTRUCTOR		AM	PM			
☐ EMPLOYEE ☐ CONTRUCTOR		AM	PM			
☐ EMPLOYEE ☐ CONTRUCTOR		AM	PM			

EQUIPMENT ON SITE	NO. OF UNITE	WORKING YES / NO

HIRED EQUIPMENT	NO. OF UNITE	EQUIPMENT RENTED	FROM	RATE

NAME: _______________________   SIGNATURE: _______________________

MO TU WE TH FR SA SU
☐   ☐   ☐   ☐   ☐   ☐   ☐

DATE: /   /

PROJECT:

FOREMAN:

WEATHER

F°____ C°____ ____AM ____PM

| HOURS DUE TO BAD WEATHER | ISSUED AND DELAYS |

NOTE: ____

COMPLETION DATE	DAYS AHEAD OF SCHEDULE	DAYS BEHIND SCHEDULE

SAFETY AND INCIDENTS

SAFETY ISSUES THAT NEED TO BE ADDRESSED	ACCIDENTS / INCIDENTS / STEPS NEEDED TO RESOLVE

SUMMARY OF THE WORK DONE TODAY

IMPORTANT NOTES

NAME	SIGNATURE

TODAY LABOR

INITIALS	TRADE	START	FINISH	PAID HOURS	OVERTIME	COMPANY
☐ EMPLOYEE ☐ CONTRUCTOR		AM	PM			
☐ EMPLOYEE ☐ CONTRUCTOR		AM	PM			
☐ EMPLOYEE ☐ CONTRUCTOR		AM	PM			
☐ EMPLOYEE ☐ CONTRUCTOR		AM	PM			
☐ EMPLOYEE ☐ CONTRUCTOR		AM	PM			
☐ EMPLOYEE ☐ CONTRUCTOR		AM	PM			
☐ EMPLOYEE ☐ CONTRUCTOR		AM	PM			
☐ EMPLOYEE ☐ CONTRUCTOR		AM	PM			

EQUIPMENT ON SITE	NO. OF UNITE	WORKING YES / NO

HIRED EQUIPMENT	NO. OF UNITE	EQUIPMENT RENTED	FROM	RATE

NAME: _______________   SIGNATURE: _______________

MO ☐ TU ☐ WE ☐ TH ☐ FR ☐ SA ☐ SU ☐

DATE: ___ / ___ / ___

PROJECT:

FOREMAN:

WEATHER F°_____ C°_____   _____ AM _____ PM

HOURS DUE TO BAD WEATHER

ISSUED AND DELAYS

NOTE: _______________________________

COMPLETION DATE	DAYS AHEAD OF SCHEDULE	DAYS BEHIND SCHEDULE

SAFETY AND INCIDENTS

SAFETY ISSUES THAT NEED TO BE ADDRESSED	ACCIDENTS / INCIDENTS / STEPS NEEDED TO RESOLVE

SUMMARY OF THE WORK DONE TODAY

IMPORTANT NOTES

NAME	SIGNATURE

TODAY LABOR

INITIALS	TRADE	START	FINISH	PAID HOURS	OVERTIME	COMPANY
☐ EMPLOYEE ☐ CONTRUCTOR		AM	PM			
☐ EMPLOYEE ☐ CONTRUCTOR		AM	PM			
☐ EMPLOYEE ☐ CONTRUCTOR		AM	PM			
☐ EMPLOYEE ☐ CONTRUCTOR		AM	PM			
☐ EMPLOYEE ☐ CONTRUCTOR		AM	PM			
☐ EMPLOYEE ☐ CONTRUCTOR		AM	PM			
☐ EMPLOYEE ☐ CONTRUCTOR		AM	PM			
☐ EMPLOYEE ☐ CONTRUCTOR		AM	PM			

EQUIPMENT ON SITE	NO. OF UNITE	WORKING YES / NO

HIRED EQUIPMENT	NO. OF UNITE	EQUIPMENT RENTED	FROM	RATE

NAME: ______________________   SIGNATURE: ______________________

MO   TU   WE   TH   FR   SA   SU
☐   ☐   ☐   ☐   ☐   ☐   ☐

DATE: _____ / _____ / _____

PROJECT:

FOREMAN:

WEATHER   F° ____   C° ____   ____ AM   ____ PM

HOURS DUE TO BAD WEATHER	ISSUED AND DELAYS

NOTE: ______________________________

COMPLETION DATE	DAYS AHEAD OF SCHEDULE	DAYS BEHIND SCHEDULE

SAFETY AND INCIDENTS

SAFETY ISSUES THAT NEED TO BE ADDRESSED	ACCIDENTS / INCIDENTS / STEPS NEEDED TO RESOLVE

SUMMARY OF THE WORK DONE TODAY

IMPORTANT NOTES

NAME	SIGNATURE

TODAY LABOR						
INITIALS	TRADE	START	FINISH	PAID HOURS	OVERTIME	COMPANY
☐ EMPLOYEE ☐ CONTRUCTOR		AM	PM			
☐ EMPLOYEE ☐ CONTRUCTOR		AM	PM			
☐ EMPLOYEE ☐ CONTRUCTOR		AM	PM			
☐ EMPLOYEE ☐ CONTRUCTOR		AM	PM			
☐ EMPLOYEE ☐ CONTRUCTOR		AM	PM			
☐ EMPLOYEE ☐ CONTRUCTOR		AM	PM			
☐ EMPLOYEE ☐ CONTRUCTOR		AM	PM			
☐ EMPLOYEE ☐ CONTRUCTOR		AM	PM			

EQUIPMENT ON SITE	NO. OF UNITE	WORKING YES / NO

HIRED EQUIPMENT	NO. OF UNITE	EQUIPMENT RENTED	FROM	RATE

NAME: SIGNATURE:

MO TU WE TH FR SA SU
☐   ☐   ☐   ☐   ☐   ☐   ☐

DATE: /   /

PROJECT:

FOREMAN:

WEATHER

F°_____ C°_____   _____ AM   _____ PM

HOURS DUE TO BAD WEATHER

ISSUED AND DELAYS

NOTE:

COMPLETION DATE	DAYS AHEAD OF SCHEDULE	DAYS BEHIND SCHEDULE

SAFETY AND INCIDENTS

SAFETY ISSUES THAT NEED TO BE ADDRESSED	ACCIDENTS / INCIDENTS / STEPS NEEDED TO RESOLVE

SUMMARY OF THE WORK DONE TODAY

IMPORTANT NOTES

NAME	SIGNATURE

TODAY LABOR							
INITIALS	TRADE	START	FINISH	PAID HOURS	OVERTIME	COMPANY
☐ EMPLOYEE ☐ CONTRUCTOR		AM	PM			
☐ EMPLOYEE ☐ CONTRUCTOR		AM	PM			
☐ EMPLOYEE ☐ CONTRUCTOR		AM	PM			
☐ EMPLOYEE ☐ CONTRUCTOR		AM	PM			
☐ EMPLOYEE ☐ CONTRUCTOR		AM	PM			
☐ EMPLOYEE ☐ CONTRUCTOR		AM	PM			
☐ EMPLOYEE ☐ CONTRUCTOR		AM	PM			
☐ EMPLOYEE ☐ CONTRUCTOR		AM	PM			

EQUIPMENT ON SITE	NO. OF UNITE	WORKING YES / NO

HIRED EQUIPMENT	NO. OF UNITE	EQUIPMENT RENTED	FROM	RATE

NAME: _______________________   SIGNATURE: _______________________

MO   TU   WE   TH   FR   SA   SU
☐   ☐   ☐   ☐   ☐   ☐   ☐

DATE: /   /

PROJECT:

FOREMAN:

WEATHER

F° _____   C° _____   AM _____   PM _____

HOURS DUE TO
BAD WEATHER

ISSUED AND DELAYS

NOTE: _______________________________________

COMPLETION DATE	DAYS AHEAD OF SCHEDULE	DAYS BEHIND SCHEDULE

SAFETY AND INCIDENTS

SAFETY ISSUES THAT NEED TO BE ADDRESSED	ACCIDENTS / INCIDENTS / STEPS NEEDED TO RESOLVE

SUMMARY OF THE WORK DONE TODAY

IMPORTANT NOTES

NAME	SIGNATURE

TODAY LABOR

INITIALS	TRADE	START	FINISH	PAID HOURS	OVERTIME	COMPANY
☐ EMPLOYEE / ☐ CONTRUCTOR		AM	PM			
☐ EMPLOYEE / ☐ CONTRUCTOR		AM	PM			
☐ EMPLOYEE / ☐ CONTRUCTOR		AM	PM			
☐ EMPLOYEE / ☐ CONTRUCTOR		AM	PM			
☐ EMPLOYEE / ☐ CONTRUCTOR		AM	PM			
☐ EMPLOYEE / ☐ CONTRUCTOR		AM	PM			
☐ EMPLOYEE / ☐ CONTRUCTOR		AM	PM			
☐ EMPLOYEE / ☐ CONTRUCTOR		AM	PM			

EQUIPMENT ON SITE	NO. OF UNITE	WORKING YES / NO

HIRED EQUIPMENT	NO. OF UNITE	EQUIPMENT RENTED	FROM	RATE

NAME: __________________________   SIGNATURE: __________________________

MO TU WE TH FR SA SU
☐   ☐   ☐   ☐   ☐   ☐   ☐

DATE: /   /

PROJECT:

FOREMAN:

WEATHER

F°_____ C°_____   _____ AM _____ PM

HOURS DUE TO
BAD WEATHER

ISSUED AND DELAYS

NOTE: _______________________________________

COMPLETION DATE	DAYS AHEAD OF SCHEDULE	DAYS BEHIND SCHEDULE

SAFETY AND INCIDENTS

SAFETY ISSUES THAT NEED TO BE ADDRESSED	ACCIDENTS / INCIDENTS / STEPS NEEDED TO RESOLVE

SUMMARY OF THE WORK DONE TODAY

IMPORTANT NOTES

NAME	SIGNATURE

TODAY LABOR

INITIALS	TRADE	START	FINISH	PAID HOURS	OVERTIME	COMPANY
☐ EMPLOYEE ☐ CONTRUCTOR		AM	PM			
☐ EMPLOYEE ☐ CONTRUCTOR		AM	PM			
☐ EMPLOYEE ☐ CONTRUCTOR		AM	PM			
☐ EMPLOYEE ☐ CONTRUCTOR		AM	PM			
☐ EMPLOYEE ☐ CONTRUCTOR		AM	PM			
☐ EMPLOYEE ☐ CONTRUCTOR		AM	PM			
☐ EMPLOYEE ☐ CONTRUCTOR		AM	PM			
☐ EMPLOYEE ☐ CONTRUCTOR		AM	PM			

EQUIPMENT ON SITE	NO. OF UNITE	WORKING YES / NO

HIRED EQUIPMENT	NO. OF UNITE	EQUIPMENT RENTED	FROM	RATE

NAME: _______________   SIGNATURE: _______________

MO   TU   WE   TH   FR   SA   SU

☐   ☐   ☐   ☐   ☐   ☐   ☐

DATE: /   /

PROJECT:

FOREMAN:

WEATHER

F°   C°   AM   PM

HOURS DUE TO BAD WEATHER

ISSUED AND DELAYS

NOTE:

COMPLETION DATE	DAYS AHEAD OF SCHEDULE	DAYS BEHIND SCHEDULE

SAFETY AND INCIDENTS

SAFETY ISSUES THAT NEED TO BE ADDRESSED	ACCIDENTS / INCIDENTS / STEPS NEEDED TO RESOLVE

SUMMARY OF THE WORK DONE TODAY

IMPORTANT NOTES

NAME	SIGNATURE

TODAY LABOR

INITIALS	TRADE	START	FINISH	PAID HOURS	OVERTIME	COMPANY
☐ EMPLOYEE ☐ CONTRUCTOR		AM	PM			
☐ EMPLOYEE ☐ CONTRUCTOR		AM	PM			
☐ EMPLOYEE ☐ CONTRUCTOR		AM	PM			
☐ EMPLOYEE ☐ CONTRUCTOR		AM	PM			
☐ EMPLOYEE ☐ CONTRUCTOR		AM	PM			
☐ EMPLOYEE ☐ CONTRUCTOR		AM	PM			
☐ EMPLOYEE ☐ CONTRUCTOR		AM	PM			
☐ EMPLOYEE ☐ CONTRUCTOR		AM	PM			

EQUIPMENT ON SITE	NO. OF UNITE	WORKING YES / NO

HIRED EQUIPMENT	NO. OF UNITE	EQUIPMENT RENTED	FROM	RATE

NAME: ____________________   SIGNATURE: ____________________

MO TU WE TH FR SA SU
☐   ☐   ☐   ☐   ☐   ☐   ☐

DATE: ___ / ___ / ___

PROJECT:

FOREMAN:

WEATHER   F°____   C°____   ____ AM   ____ PM

HOURS DUE TO BAD WEATHER

ISSUED AND DELAYS

NOTE: ____

COMPLETION DATE	DAYS AHEAD OF SCHEDULE	DAYS BEHIND SCHEDULE

SAFETY AND INCIDENTS

SAFETY ISSUES THAT NEED TO BE ADDRESSED	ACCIDENTS / INCIDENTS / STEPS NEEDED TO RESOLVE

SUMMARY OF THE WORK DONE TODAY

IMPORTANT NOTES

NAME	SIGNATURE

TODAY LABOR							
INITIALS	TRADE	START	FINISH	PAID HOURS	OVERTIME	COMPANY
☐ EMPLOYEE ☐ CONTRUCTOR		AM	PM			
☐ EMPLOYEE ☐ CONTRUCTOR		AM	PM			
☐ EMPLOYEE ☐ CONTRUCTOR		AM	PM			
☐ EMPLOYEE ☐ CONTRUCTOR		AM	PM			
☐ EMPLOYEE ☐ CONTRUCTOR		AM	PM			
☐ EMPLOYEE ☐ CONTRUCTOR		AM	PM			
☐ EMPLOYEE ☐ CONTRUCTOR		AM	PM			
☐ EMPLOYEE ☐ CONTRUCTOR		AM	PM			

EQUIPMENT ON SITE	NO. OF UNITE	WORKING YES / NO

HIRED EQUIPMENT	NO. OF UNITE	EQUIPMENT RENTED	FROM	RATE

NAME: _______________________   SIGNATURE: _______________________

MO	TU	WE	TH	FR	SA	SU		DATE: / /

PROJECT:

FOREMAN:

WEATHER F°_____ C°_____ _____ AM _____ PM

| HOURS DUE TO BAD WEATHER | ISSUED AND DELAYS |

NOTE: ________________________________

COMPLETION DATE	DAYS AHEAD OF SCHEDULE	DAYS BEHIND SCHEDULE

SAFETY AND INCIDENTS

SAFETY ISSUES THAT NEED TO BE ADDRESSED	ACCIDENTS / INCIDENTS / STEPS NEEDED TO RESOLVE

SUMMARY OF THE WORK DONE TODAY

IMPORTANT NOTES

NAME	SIGNATURE

TODAY LABOR

INITIALS	TRADE	START	FINISH	PAID HOURS	OVERTIME	COMPANY
☐ EMPLOYEE ☐ CONTRUCTOR		AM	PM			
☐ EMPLOYEE ☐ CONTRUCTOR		AM	PM			
☐ EMPLOYEE ☐ CONTRUCTOR		AM	PM			
☐ EMPLOYEE ☐ CONTRUCTOR		AM	PM			
☐ EMPLOYEE ☐ CONTRUCTOR		AM	PM			
☐ EMPLOYEE ☐ CONTRUCTOR		AM	PM			
☐ EMPLOYEE ☐ CONTRUCTOR		AM	PM			
☐ EMPLOYEE ☐ CONTRUCTOR		AM	PM			

EQUIPMENT ON SITE	NO. OF UNITE	WORKING YES / NO

HIRED EQUIPMENT	NO. OF UNITE	EQUIPMENT RENTED	FROM	RATE

NAME: _______________________   SIGNATURE: _______________________

MO TU WE TH FR SA SU
☐   ☐   ☐   ☐   ☐   ☐   ☐

DATE: /   /

PROJECT:

FOREMAN:

WEATHER

F°____ C°____   ____ AM   ____ PM

HOURS DUE TO BAD WEATHER

ISSUED AND DELAYS

NOTE: ____

COMPLETION DATE	DAYS AHEAD OF SCHEDULE	DAYS BEHIND SCHEDULE

SAFETY AND INCIDENTS

SAFETY ISSUES THAT NEED TO BE ADDRESSED	ACCIDENTS / INCIDENTS / STEPS NEEDED TO RESOLVE

SUMMARY OF THE WORK DONE TODAY

IMPORTANT NOTES

NAME	SIGNATURE

<table>
<tr><td colspan="7" align="center">TODAY LABOR</td></tr>
</table>

INITIALS	TRADE	START	FINISH	PAID HOURS	OVERTIME	COMPANY
☐ EMPLOYEE ☐ CONTRUCTOR		AM	PM			
☐ EMPLOYEE ☐ CONTRUCTOR		AM	PM			
☐ EMPLOYEE ☐ CONTRUCTOR		AM	PM			
☐ EMPLOYEE ☐ CONTRUCTOR		AM	PM			
☐ EMPLOYEE ☐ CONTRUCTOR		AM	PM			
☐ EMPLOYEE ☐ CONTRUCTOR		AM	PM			
☐ EMPLOYEE ☐ CONTRUCTOR		AM	PM			
☐ EMPLOYEE ☐ CONTRUCTOR		AM	PM			

EQUIPMENT ON SITE	NO. OF UNITE	WORKING YES / NO

HIRED EQUIPMENT	NO. OF UNITE	EQUIPMENT RENTED	FROM	RATE

NAME: ____________________ SIGNATURE: ____________________

DATE: /   /

PROJECT:

FOREMAN:

WEATHER

F°   C°   AM   PM

HOURS DUE TO BAD WEATHER

ISSUED AND DELAYS

NOTE:

COMPLETION DATE	DAYS AHEAD OF SCHEDULE	DAYS BEHIND SCHEDULE

SAFETY AND INCIDENTS

SAFETY ISSUES THAT NEED TO BE ADDRESSED	ACCIDENTS / INCIDENTS / STEPS NEEDED TO RESOLVE

SUMMARY OF THE WORK DONE TODAY

IMPORTANT NOTES

NAME	SIGNATURE

TODAY LABOR						
INITIALS	TRADE	START	FINISH	PAID HOURS	OVERTIME	COMPANY
☐ EMPLOYEE ☐ CONTRUCTOR		AM	PM			
☐ EMPLOYEE ☐ CONTRUCTOR		AM	PM			
☐ EMPLOYEE ☐ CONTRUCTOR		AM	PM			
☐ EMPLOYEE ☐ CONTRUCTOR		AM	PM			
☐ EMPLOYEE ☐ CONTRUCTOR		AM	PM			
☐ EMPLOYEE ☐ CONTRUCTOR		AM	PM			
☐ EMPLOYEE ☐ CONTRUCTOR		AM	PM			
☐ EMPLOYEE ☐ CONTRUCTOR		AM	PM			

EQUIPMENT ON SITE	NO. OF UNITE	WORKING YES / NO

HIRED EQUIPMENT	NO. OF UNITE	EQUIPMENT RENTED	FROM	RATE

NAME: ______________________   SIGNATURE: ______________________

MO TU WE TH FR SA SU
☐   ☐   ☐   ☐   ☐   ☐   ☐

DATE: /   /

PROJECT:

FOREMAN:

WEATHER

F°_____ C°_____   AM_____   PM_____

HOURS DUE TO BAD WEATHER	ISSUED AND DELAYS

NOTE: _______________________

COMPLETION DATE	DAYS AHEAD OF SCHEDULE	DAYS BEHIND SCHEDULE

SAFETY AND INCIDENTS

SAFETY ISSUES THAT NEED TO BE ADDRESSED	ACCIDENTS / INCIDENTS / STEPS NEEDED TO RESOLVE

SUMMARY OF THE WORK DONE TODAY

IMPORTANT NOTES

NAME	SIGNATURE

<table>
<thead>
<tr><th colspan="8" align="center">TODAY LABOR</th></tr>
<tr><th>INITIALS</th><th>TRADE</th><th>START</th><th>FINISH</th><th>PAID HOURS</th><th>OVERTIME</th><th>COMPANY</th><th></th></tr>
</thead>
<tbody>
<tr><td>☐ EMPLOYEE
☐ CONTRUCTOR</td><td></td><td>AM</td><td>PM</td><td></td><td></td><td></td><td></td></tr>
<tr><td>☐ EMPLOYEE
☐ CONTRUCTOR</td><td></td><td>AM</td><td>PM</td><td></td><td></td><td></td><td></td></tr>
<tr><td>☐ EMPLOYEE
☐ CONTRUCTOR</td><td></td><td>AM</td><td>PM</td><td></td><td></td><td></td><td></td></tr>
<tr><td>☐ EMPLOYEE
☐ CONTRUCTOR</td><td></td><td>AM</td><td>PM</td><td></td><td></td><td></td><td></td></tr>
<tr><td>☐ EMPLOYEE
☐ CONTRUCTOR</td><td></td><td>AM</td><td>PM</td><td></td><td></td><td></td><td></td></tr>
<tr><td>☐ EMPLOYEE
☐ CONTRUCTOR</td><td></td><td>AM</td><td>PM</td><td></td><td></td><td></td><td></td></tr>
<tr><td>☐ EMPLOYEE
☐ CONTRUCTOR</td><td></td><td>AM</td><td>PM</td><td></td><td></td><td></td><td></td></tr>
<tr><td>☐ EMPLOYEE
☐ CONTRUCTOR</td><td></td><td>AM</td><td>PM</td><td></td><td></td><td></td><td></td></tr>
</tbody>
</table>

EQUIPMENT ON SITE	NO. OF UNITE	WORKING YES / NO

HIRED EQUIPMENT	NO. OF UNITE	EQUIPMENT RENTED	FROM	RATE

NAME: SIGNATURE:

MO   TU   WE   TH   FR   SA   SU
DATE: /   /

PROJECT:

FOREMAN:

WEATHER   F° ___ C° ___   ___ AM   ___ PM

HOURS DUE TO BAD WEATHER	ISSUED AND DELAYS

NOTE: ____

COMPLETION DATE	DAYS AHEAD OF SCHEDULE	DAYS BEHIND SCHEDULE

SAFETY AND INCIDENTS

SAFETY ISSUES THAT NEED TO BE ADDRESSED	ACCIDENTS / INCIDENTS / STEPS NEEDED TO RESOLVE

SUMMARY OF THE WORK DONE TODAY

IMPORTANT NOTES

NAME	SIGNATURE

TODAY LABOR

INITIALS	TRADE	START	FINISH	PAID HOURS	OVERTIME	COMPANY
☐ EMPLOYEE ☐ CONTRUCTOR		AM	PM			
☐ EMPLOYEE ☐ CONTRUCTOR		AM	PM			
☐ EMPLOYEE ☐ CONTRUCTOR		AM	PM			
☐ EMPLOYEE ☐ CONTRUCTOR		AM	PM			
☐ EMPLOYEE ☐ CONTRUCTOR		AM	PM			
☐ EMPLOYEE ☐ CONTRUCTOR		AM	PM			
☐ EMPLOYEE ☐ CONTRUCTOR		AM	PM			
☐ EMPLOYEE ☐ CONTRUCTOR		AM	PM			

EQUIPMENT ON SITE	NO. OF UNITE	WORKING YES / NO

HIRED EQUIPMENT	NO. OF UNITE	EQUIPMENT RENTED	FROM	RATE

NAME: _____________________   SIGNATURE: _____________________

MO TU WE TH FR SA SU
☐   ☐   ☐   ☐   ☐   ☐   ☐

DATE: /   /

PROJECT:

FOREMAN:

WEATHER   F°____ C°____   ____ AM ____ PM

HOURS DUE TO BAD WEATHER

ISSUED AND DELAYS

NOTE:

COMPLETION DATE	DAYS AHEAD OF SCHEDULE	DAYS BEHIND SCHEDULE

SAFETY AND INCIDENTS

SAFETY ISSUES THAT NEED TO BE ADDRESSED	ACCIDENTS / INCIDENTS / STEPS NEEDED TO RESOLVE

SUMMARY OF THE WORK DONE TODAY

IMPORTANT NOTES

NAME	SIGNATURE

<table>
<tr><td colspan="8" align="center">TODAY LABOR</td></tr>
<tr><td>INITIALS</td><td>TRADE</td><td>START</td><td>FINISH</td><td>PAID HOURS</td><td>OVERTIME</td><td colspan="2">COMPANY</td></tr>
<tr><td>☐ EMPLOYEE
☐ CONTRUCTOR</td><td></td><td>AM</td><td>PM</td><td></td><td></td><td colspan="2"></td></tr>
<tr><td>☐ EMPLOYEE
☐ CONTRUCTOR</td><td></td><td>AM</td><td>PM</td><td></td><td></td><td colspan="2"></td></tr>
<tr><td>☐ EMPLOYEE
☐ CONTRUCTOR</td><td></td><td>AM</td><td>PM</td><td></td><td></td><td colspan="2"></td></tr>
<tr><td>☐ EMPLOYEE
☐ CONTRUCTOR</td><td></td><td>AM</td><td>PM</td><td></td><td></td><td colspan="2"></td></tr>
<tr><td>☐ EMPLOYEE
☐ CONTRUCTOR</td><td></td><td>AM</td><td>PM</td><td></td><td></td><td colspan="2"></td></tr>
<tr><td>☐ EMPLOYEE
☐ CONTRUCTOR</td><td></td><td>AM</td><td>PM</td><td></td><td></td><td colspan="2"></td></tr>
<tr><td>☐ EMPLOYEE
☐ CONTRUCTOR</td><td></td><td>AM</td><td>PM</td><td></td><td></td><td colspan="2"></td></tr>
<tr><td>☐ EMPLOYEE
☐ CONTRUCTOR</td><td></td><td>AM</td><td>PM</td><td></td><td></td><td colspan="2"></td></tr>
</table>

EQUIPMENT ON SITE	NO. OF UNITE	WORKING YES / NO

HIRED EQUIPMENT	NO. OF UNITE	EQUIPMENT RENTED	FROM	RATE

NAME: _______________________   SIGNATURE: _______________________

MO   TU   WE   TH   FR   SA   SU
☐   ☐   ☐   ☐   ☐   ☐   ☐

DATE: /   /

PROJECT:

FOREMAN:

WEATHER

F°______ C°______   ______ AM   ______ PM

HOURS DUE TO
BAD WEATHER

ISSUED AND DELAYS

NOTE:

COMPLETION DATE	DAYS AHEAD OF SCHEDULE	DAYS BEHIND SCHEDULE

SAFETY AND INCIDENTS

SAFETY ISSUES THAT NEED TO BE ADDRESSED	ACCIDENTS / INCIDENTS / STEPS NEEDED TO RESOLVE

SUMMARY OF THE WORK DONE TODAY

IMPORTANT NOTES

NAME	SIGNATURE

TODAY LABOR

INITIALS	TRADE	START	FINISH	PAID HOURS	OVERTIME	COMPANY
☐ EMPLOYEE ☐ CONTRUCTOR		AM	PM			
☐ EMPLOYEE ☐ CONTRUCTOR		AM	PM			
☐ EMPLOYEE ☐ CONTRUCTOR		AM	PM			
☐ EMPLOYEE ☐ CONTRUCTOR		AM	PM			
☐ EMPLOYEE ☐ CONTRUCTOR		AM	PM			
☐ EMPLOYEE ☐ CONTRUCTOR		AM	PM			
☐ EMPLOYEE ☐ CONTRUCTOR		AM	PM			
☐ EMPLOYEE ☐ CONTRUCTOR		AM	PM			

EQUIPMENT ON SITE	NO. OF UNITE	WORKING YES / NO

HIRED EQUIPMENT	NO. OF UNITE	EQUIPMENT RENTED	FROM	RATE

NAME: _______________________   SIGNATURE: _______________________

MO TU WE TH FR SA SU
☐   ☐   ☐   ☐   ☐   ☐   ☐

DATE: /   /

PROJECT:

FOREMAN:

WEATHER   F°____ C°____   ____ AM ____ PM

HOURS DUE TO BAD WEATHER

ISSUED AND DELAYS

NOTE: _______________________________________

COMPLETION DATE	DAYS AHEAD OF SCHEDULE	DAYS BEHIND SCHEDULE

SAFETY AND INCIDENTS

SAFETY ISSUES THAT NEED TO BE ADDRESSED	ACCIDENTS / INCIDENTS / STEPS NEEDED TO RESOLVE

SUMMARY OF THE WORK DONE TODAY

IMPORTANT NOTES

NAME	SIGNATURE

TODAY LABOR

INITIALS	TRADE	START	FINISH	PAID HOURS	OVERTIME	COMPANY
☐ EMPLOYEE ☐ CONTRUCTOR		AM	PM			
☐ EMPLOYEE ☐ CONTRUCTOR		AM	PM			
☐ EMPLOYEE ☐ CONTRUCTOR		AM	PM			
☐ EMPLOYEE ☐ CONTRUCTOR		AM	PM			
☐ EMPLOYEE ☐ CONTRUCTOR		AM	PM			
☐ EMPLOYEE ☐ CONTRUCTOR		AM	PM			
☐ EMPLOYEE ☐ CONTRUCTOR		AM	PM			
☐ EMPLOYEE ☐ CONTRUCTOR		AM	PM			

EQUIPMENT ON SITE	NO. OF UNITE	WORKING YES / NO

HIRED EQUIPMENT	NO. OF UNITE	EQUIPMENT RENTED	FROM	RATE

NAME: _______________________   SIGNATURE: _______________________

MO TU WE TH FR SA SU
☐   ☐   ☐   ☐   ☐   ☐   ☐

DATE: /   /

PROJECT:

FOREMAN:

WEATHER

F°____ C°____   ____ AM   ____ PM

HOURS DUE TO BAD WEATHER

ISSUED AND DELAYS

NOTE: ____________________

COMPLETION DATE	DAYS AHEAD OF SCHEDULE	DAYS BEHIND SCHEDULE

SAFETY AND INCIDENTS

SAFETY ISSUES THAT NEED TO BE ADDRESSED	ACCIDENTS / INCIDENTS / STEPS NEEDED TO RESOLVE

SUMMARY OF THE WORK DONE TODAY

IMPORTANT NOTES

NAME	SIGNATURE

TODAY LABOR						
INITIALS	TRADE	START	FINISH	PAID HOURS	OVERTIME	COMPANY
☐ EMPLOYEE ☐ CONTRUCTOR		AM	PM			
☐ EMPLOYEE ☐ CONTRUCTOR		AM	PM			
☐ EMPLOYEE ☐ CONTRUCTOR		AM	PM			
☐ EMPLOYEE ☐ CONTRUCTOR		AM	PM			
☐ EMPLOYEE ☐ CONTRUCTOR		AM	PM			
☐ EMPLOYEE ☐ CONTRUCTOR		AM	PM			
☐ EMPLOYEE ☐ CONTRUCTOR		AM	PM			
☐ EMPLOYEE ☐ CONTRUCTOR		AM	PM			

EQUIPMENT ON SITE	NO. OF UNITE	WORKING YES / NO

HIRED EQUIPMENT	NO. OF UNITE	EQUIPMENT RENTED	FROM	RATE

NAME: ___________________________   SIGNATURE: ___________________________

MO   TU   WE   TH   FR   SA   SU

DATE: / /

PROJECT:

FOREMAN:

WEATHER

F°   C°   AM   PM

HOURS DUE TO BAD WEATHER

ISSUED AND DELAYS

NOTE:

COMPLETION DATE	DAYS AHEAD OF SCHEDULE	DAYS BEHIND SCHEDULE

SAFETY AND INCIDENTS

SAFETY ISSUES THAT NEED TO BE ADDRESSED	ACCIDENTS / INCIDENTS / STEPS NEEDED TO RESOLVE

SUMMARY OF THE WORK DONE TODAY

IMPORTANT NOTES

NAME	SIGNATURE

TODAY LABOR

INITIALS	TRADE	START	FINISH	PAID HOURS	OVERTIME	COMPANY
☐ EMPLOYEE ☐ CONTRUCTOR		AM	PM			
☐ EMPLOYEE ☐ CONTRUCTOR		AM	PM			
☐ EMPLOYEE ☐ CONTRUCTOR		AM	PM			
☐ EMPLOYEE ☐ CONTRUCTOR		AM	PM			
☐ EMPLOYEE ☐ CONTRUCTOR		AM	PM			
☐ EMPLOYEE ☐ CONTRUCTOR		AM	PM			
☐ EMPLOYEE ☐ CONTRUCTOR		AM	PM			
☐ EMPLOYEE ☐ CONTRUCTOR		AM	PM			

EQUIPMENT ON SITE	NO. OF UNITE	WORKING YES / NO

HIRED EQUIPMENT	NO. OF UNITE	EQUIPMENT RENTED	FROM	RATE

NAME: _______________________   SIGNATURE: _______________________

| MO | TU | WE | TH | FR | SA | SU | | DATE: /   / |
| ☐ | ☐ | ☐ | ☐ | ☐ | ☐ | ☐ | | |

PROJECT:

FOREMAN:

WEATHER

F° ___ C° ___ ___ AM ___ PM

HOURS DUE TO BAD WEATHER	ISSUED AND DELAYS

NOTE: ___________________

COMPLETION DATE	DAYS AHEAD OF SCHEDULE	DAYS BEHIND SCHEDULE

SAFETY AND INCIDENTS

SAFETY ISSUES THAT NEED TO BE ADDRESSED	ACCIDENTS / INCIDENTS / STEPS NEEDED TO RESOLVE

SUMMARY OF THE WORK DONE TODAY

IMPORTANT NOTES

NAME	SIGNATURE

TODAY LABOR

INITIALS	TRADE	START	FINISH	PAID HOURS	OVERTIME	COMPANY
☐ EMPLOYEE ☐ CONTRUCTOR		AM	PM			
☐ EMPLOYEE ☐ CONTRUCTOR		AM	PM			
☐ EMPLOYEE ☐ CONTRUCTOR		AM	PM			
☐ EMPLOYEE ☐ CONTRUCTOR		AM	PM			
☐ EMPLOYEE ☐ CONTRUCTOR		AM	PM			
☐ EMPLOYEE ☐ CONTRUCTOR		AM	PM			
☐ EMPLOYEE ☐ CONTRUCTOR		AM	PM			
☐ EMPLOYEE ☐ CONTRUCTOR		AM	PM			

EQUIPMENT ON SITE	NO. OF UNITE	WORKING YES / NO

HIRED EQUIPMENT	NO. OF UNITE	EQUIPMENT RENTED	FROM	RATE

NAME: SIGNATURE:

MO   TU   WE   TH   FR   SA   SU

DATE: /   /

PROJECT:

FOREMAN:

WEATHER

F° _____ C° _____   _____ AM _____ PM

HOURS DUE TO BAD WEATHER	ISSUED AND DELAYS

NOTE: ________________________

COMPLETION DATE	DAYS AHEAD OF SCHEDULE	DAYS BEHIND SCHEDULE

SAFETY AND INCIDENTS

SAFETY ISSUES THAT NEED TO BE ADDRESSED	ACCIDENTS / INCIDENTS / STEPS NEEDED TO RESOLVE

SUMMARY OF THE WORK DONE TODAY

IMPORTANT NOTES

NAME	SIGNATURE

TODAY LABOR

INITIALS	TRADE	START	FINISH	PAID HOURS	OVERTIME	COMPANY
☐ EMPLOYEE ☐ CONTRUCTOR		AM	PM			
☐ EMPLOYEE ☐ CONTRUCTOR		AM	PM			
☐ EMPLOYEE ☐ CONTRUCTOR		AM	PM			
☐ EMPLOYEE ☐ CONTRUCTOR		AM	PM			
☐ EMPLOYEE ☐ CONTRUCTOR		AM	PM			
☐ EMPLOYEE ☐ CONTRUCTOR		AM	PM			
☐ EMPLOYEE ☐ CONTRUCTOR		AM	PM			
☐ EMPLOYEE ☐ CONTRUCTOR		AM	PM			

EQUIPMENT ON SITE	NO. OF UNITE	WORKING YES / NO

HIRED EQUIPMENT	NO. OF UNITE	EQUIPMENT RENTED	FROM	RATE

NAME:

SIGNATURE:

MO TU WE TH FR SA SU

DATE: / /

PROJECT:

FOREMAN:

WEATHER

F°   C°   AM   PM

HOURS DUE TO BAD WEATHER

ISSUED AND DELAYS

NOTE:

COMPLETION DATE	DAYS AHEAD OF SCHEDULE	DAYS BEHIND SCHEDULE

SAFETY AND INCIDENTS

SAFETY ISSUES THAT NEED TO BE ADDRESSED	ACCIDENTS / INCIDENTS / STEPS NEEDED TO RESOLVE

SUMMARY OF THE WORK DONE TODAY

IMPORTANT NOTES

NAME	SIGNATURE

TODAY LABOR

INITIALS	TRADE	START	FINISH	PAID HOURS	OVERTIME	COMPANY
☐ EMPLOYEE ☐ CONTRUCTOR		AM	PM			
☐ EMPLOYEE ☐ CONTRUCTOR		AM	PM			
☐ EMPLOYEE ☐ CONTRUCTOR		AM	PM			
☐ EMPLOYEE ☐ CONTRUCTOR		AM	PM			
☐ EMPLOYEE ☐ CONTRUCTOR		AM	PM			
☐ EMPLOYEE ☐ CONTRUCTOR		AM	PM			
☐ EMPLOYEE ☐ CONTRUCTOR		AM	PM			
☐ EMPLOYEE ☐ CONTRUCTOR		AM	PM			

EQUIPMENT ON SITE	NO. OF UNITE	WORKING YES / NO

HIRED EQUIPMENT	NO. OF UNITE	EQUIPMENT RENTED	FROM	RATE

NAME: _______________________   SIGNATURE: _______________________

MO TU WE TH FR SA SU
☐   ☐   ☐   ☐   ☐   ☐   ☐

DATE: /   /

PROJECT:

FOREMAN:

WEATHER   F°_____ C°_____   _____ AM _____ PM

HOURS DUE TO BAD WEATHER

ISSUED AND DELAYS

NOTE: ____

COMPLETION DATE	DAYS AHEAD OF SCHEDULE	DAYS BEHIND SCHEDULE

SAFETY AND INCIDENTS

SAFETY ISSUES THAT NEED TO BE ADDRESSED	ACCIDENTS / INCIDENTS / STEPS NEEDED TO RESOLVE

SUMMARY OF THE WORK DONE TODAY

IMPORTANT NOTES

NAME	SIGNATURE

TODAY LABOR

INITIALS	TRADE	START	FINISH	PAID HOURS	OVERTIME	COMPANY
☐ EMPLOYEE ☐ CONTRUCTOR		AM	PM			
☐ EMPLOYEE ☐ CONTRUCTOR		AM	PM			
☐ EMPLOYEE ☐ CONTRUCTOR		AM	PM			
☐ EMPLOYEE ☐ CONTRUCTOR		AM	PM			
☐ EMPLOYEE ☐ CONTRUCTOR		AM	PM			
☐ EMPLOYEE ☐ CONTRUCTOR		AM	PM			
☐ EMPLOYEE ☐ CONTRUCTOR		AM	PM			
☐ EMPLOYEE ☐ CONTRUCTOR		AM	PM			

EQUIPMENT ON SITE	NO. OF UNITE	WORKING YES / NO

HIRED EQUIPMENT	NO. OF UNITE	EQUIPMENT RENTED	FROM	RATE

NAME: ___________________   SIGNATURE: ___________________

MO ☐ TU ☐ WE ☐ TH ☐ FR ☐ SA ☐ SU ☐

DATE: ___ / ___ / ___

PROJECT:

FOREMAN:

WEATHER

F° ___ C° ___ ___ AM ___ PM

HOURS DUE TO BAD WEATHER

ISSUED AND DELAYS

NOTE:

COMPLETION DATE	DAYS AHEAD OF SCHEDULE	DAYS BEHIND SCHEDULE

SAFETY AND INCIDENTS

SAFETY ISSUES THAT NEED TO BE ADDRESSED	ACCIDENTS / INCIDENTS / STEPS NEEDED TO RESOLVE

SUMMARY OF THE WORK DONE TODAY

IMPORTANT NOTES

NAME	SIGNATURE

TODAY LABOR

INITIALS	TRADE	START	FINISH	PAID HOURS	OVERTIME	COMPANY
☐ EMPLOYEE ☐ CONTRUCTOR		AM	PM			
☐ EMPLOYEE ☐ CONTRUCTOR		AM	PM			
☐ EMPLOYEE ☐ CONTRUCTOR		AM	PM			
☐ EMPLOYEE ☐ CONTRUCTOR		AM	PM			
☐ EMPLOYEE ☐ CONTRUCTOR		AM	PM			
☐ EMPLOYEE ☐ CONTRUCTOR		AM	PM			
☐ EMPLOYEE ☐ CONTRUCTOR		AM	PM			
☐ EMPLOYEE ☐ CONTRUCTOR		AM	PM			

EQUIPMENT ON SITE	NO. OF UNITE	WORKING YES / NO

HIRED EQUIPMENT	NO. OF UNITE	EQUIPMENT RENTED	FROM	RATE

NAME: ________________________   SIGNATURE: ________________________

MO TU WE TH FR SA SU
☐   ☐   ☐   ☐   ☐   ☐   ☐

DATE: /   /

PROJECT:

FOREMAN:

WEATHER

F°_____ C°_____   _____ AM _____ PM

| HOURS DUE TO BAD WEATHER | ISSUED AND DELAYS |

NOTE: ____

| COMPLETION DATE | DAYS AHEAD OF SCHEDULE | DAYS BEHIND SCHEDULE |

SAFETY AND INCIDENTS

| SAFETY ISSUES THAT NEED TO BE ADDRESSED | ACCIDENTS / INCIDENTS / STEPS NEEDED TO RESOLVE |

SUMMARY OF THE WORK DONE TODAY

IMPORTANT NOTES

| NAME | SIGNATURE |

TODAY LABOR

INITIALS	TRADE	START	FINISH	PAID HOURS	OVERTIME	COMPANY
☐ EMPLOYEE ☐ CONTRUCTOR		AM	PM			
☐ EMPLOYEE ☐ CONTRUCTOR		AM	PM			
☐ EMPLOYEE ☐ CONTRUCTOR		AM	PM			
☐ EMPLOYEE ☐ CONTRUCTOR		AM	PM			
☐ EMPLOYEE ☐ CONTRUCTOR		AM	PM			
☐ EMPLOYEE ☐ CONTRUCTOR		AM	PM			
☐ EMPLOYEE ☐ CONTRUCTOR		AM	PM			
☐ EMPLOYEE ☐ CONTRUCTOR		AM	PM			

EQUIPMENT ON SITE	NO. OF UNITE	WORKING YES / NO

HIRED EQUIPMENT	NO. OF UNITE	EQUIPMENT RENTED	FROM	RATE

NAME: ______________________

SIGNATURE: ______________________

MO TU WE TH FR SA SU
☐   ☐   ☐   ☐   ☐   ☐   ☐

DATE: /   /

PROJECT:

FOREMAN:

WEATHER

F°____ C°____   ____ AM   ____ PM

HOURS DUE TO
BAD WEATHER

ISSUED AND DELAYS

NOTE:

COMPLETION DATE	DAYS AHEAD OF SCHEDULE	DAYS BEHIND SCHEDULE

SAFETY AND INCIDENTS

SAFETY ISSUES THAT NEED TO BE ADDRESSED	ACCIDENTS / INCIDENTS / STEPS NEEDED TO RESOLVE

SUMMARY OF THE WORK DONE TODAY

IMPORTANT NOTES

NAME	SIGNATURE

TODAY LABOR

INITIALS	TRADE	START	FINISH	PAID HOURS	OVERTIME	COMPANY
☐ EMPLOYEE ☐ CONTRUCTOR		AM	PM			
☐ EMPLOYEE ☐ CONTRUCTOR		AM	PM			
☐ EMPLOYEE ☐ CONTRUCTOR		AM	PM			
☐ EMPLOYEE ☐ CONTRUCTOR		AM	PM			
☐ EMPLOYEE ☐ CONTRUCTOR		AM	PM			
☐ EMPLOYEE ☐ CONTRUCTOR		AM	PM			
☐ EMPLOYEE ☐ CONTRUCTOR		AM	PM			
☐ EMPLOYEE ☐ CONTRUCTOR		AM	PM			

EQUIPMENT ON SITE	NO. OF UNITE	WORKING YES / NO

HIRED EQUIPMENT	NO. OF UNITE	EQUIPMENT RENTED	FROM	RATE

NAME: _______________________   SIGNATURE: _______________________

MO TU WE TH FR SA SU
☐   ☐   ☐   ☐   ☐   ☐   ☐

DATE: __ / __ / __

PROJECT:

FOREMAN:

WEATHER

F° _____ C° _____   _____ AM _____ PM

HOURS DUE TO BAD WEATHER	ISSUED AND DELAYS

NOTE: ____

COMPLETION DATE	DAYS AHEAD OF SCHEDULE	DAYS BEHIND SCHEDULE

SAFETY AND INCIDENTS

SAFETY ISSUES THAT NEED TO BE ADDRESSED	ACCIDENTS / INCIDENTS / STEPS NEEDED TO RESOLVE

SUMMARY OF THE WORK DONE TODAY

IMPORTANT NOTES

NAME	SIGNATURE

TODAY LABOR

INITIALS	TRADE	START	FINISH	PAID HOURS	OVERTIME	COMPANY
☐ EMPLOYEE ☐ CONTRUCTOR		AM	PM			
☐ EMPLOYEE ☐ CONTRUCTOR		AM	PM			
☐ EMPLOYEE ☐ CONTRUCTOR		AM	PM			
☐ EMPLOYEE ☐ CONTRUCTOR		AM	PM			
☐ EMPLOYEE ☐ CONTRUCTOR		AM	PM			
☐ EMPLOYEE ☐ CONTRUCTOR		AM	PM			
☐ EMPLOYEE ☐ CONTRUCTOR		AM	PM			
☐ EMPLOYEE ☐ CONTRUCTOR		AM	PM			

EQUIPMENT ON SITE	NO. OF UNITE	WORKING YES / NO

HIRED EQUIPMENT	NO. OF UNITE	EQUIPMENT RENTED	FROM	RATE

NAME: _______________________   SIGNATURE: _______________________

MO TU WE TH FR SA SU

DATE: / /

PROJECT:

FOREMAN:

WEATHER

F° C° AM PM

HOURS DUE TO BAD WEATHER

ISSUED AND DELAYS

NOTE:

COMPLETION DATE	DAYS AHEAD OF SCHEDULE	DAYS BEHIND SCHEDULE

SAFETY AND INCIDENTS

SAFETY ISSUES THAT NEED TO BE ADDRESSED	ACCIDENTS / INCIDENTS / STEPS NEEDED TO RESOLVE

SUMMARY OF THE WORK DONE TODAY

IMPORTANT NOTES

NAME	SIGNATURE

<table>
<tr><td colspan="7" align="center">TODAY LABOR</td></tr>
<tr><td>INITIALS</td><td>TRADE</td><td>START</td><td>FINISH</td><td>PAID HOURS</td><td>OVERTIME</td><td>COMPANY</td></tr>
<tr><td>☐ EMPLOYEE
☐ CONTRUCTOR</td><td></td><td>AM</td><td>PM</td><td></td><td></td><td></td></tr>
<tr><td>☐ EMPLOYEE
☐ CONTRUCTOR</td><td></td><td>AM</td><td>PM</td><td></td><td></td><td></td></tr>
<tr><td>☐ EMPLOYEE
☐ CONTRUCTOR</td><td></td><td>AM</td><td>PM</td><td></td><td></td><td></td></tr>
<tr><td>☐ EMPLOYEE
☐ CONTRUCTOR</td><td></td><td>AM</td><td>PM</td><td></td><td></td><td></td></tr>
<tr><td>☐ EMPLOYEE
☐ CONTRUCTOR</td><td></td><td>AM</td><td>PM</td><td></td><td></td><td></td></tr>
<tr><td>☐ EMPLOYEE
☐ CONTRUCTOR</td><td></td><td>AM</td><td>PM</td><td></td><td></td><td></td></tr>
<tr><td>☐ EMPLOYEE
☐ CONTRUCTOR</td><td></td><td>AM</td><td>PM</td><td></td><td></td><td></td></tr>
<tr><td>☐ EMPLOYEE
☐ CONTRUCTOR</td><td></td><td>AM</td><td>PM</td><td></td><td></td><td></td></tr>
</table>

EQUIPMENT ON SITE	NO. OF UNITE	WORKING YES / NO

HIRED EQUIPMENT	NO. OF UNITE	EQUIPMENT RENTED	FROM	RATE

NAME: SIGNATURE:

MO   TU   WE   TH   FR   SA   SU
☐   ☐   ☐   ☐   ☐   ☐   ☐

DATE: /   /

PROJECT:

FOREMAN:

WEATHER

F° ___   C° ___   ___ AM   ___ PM

| HOURS DUE TO BAD WEATHER | ISSUED AND DELAYS |

NOTE: ___________________________

COMPLETION DATE	DAYS AHEAD OF SCHEDULE	DAYS BEHIND SCHEDULE

SAFETY AND INCIDENTS

SAFETY ISSUES THAT NEED TO BE ADDRESSED	ACCIDENTS / INCIDENTS / STEPS NEEDED TO RESOLVE

SUMMARY OF THE WORK DONE TODAY

IMPORTANT NOTES

NAME	SIGNATURE

TODAY LABOR

INITIALS	TRADE	START	FINISH	PAID HOURS	OVERTIME	COMPANY
☐ EMPLOYEE ☐ CONTRUCTOR		AM	PM			
☐ EMPLOYEE ☐ CONTRUCTOR		AM	PM			
☐ EMPLOYEE ☐ CONTRUCTOR		AM	PM			
☐ EMPLOYEE ☐ CONTRUCTOR		AM	PM			
☐ EMPLOYEE ☐ CONTRUCTOR		AM	PM			
☐ EMPLOYEE ☐ CONTRUCTOR		AM	PM			
☐ EMPLOYEE ☐ CONTRUCTOR		AM	PM			
☐ EMPLOYEE ☐ CONTRUCTOR		AM	PM			

EQUIPMENT ON SITE	NO. OF UNITE	WORKING YES / NO

HIRED EQUIPMENT	NO. OF UNITE	EQUIPMENT RENTED	FROM	RATE

NAME: _______________________________   SIGNATURE: _______________________________

MO TU WE TH FR SA SU
☐   ☐   ☐   ☐   ☐   ☐   ☐

DATE: /   /

PROJECT:

FOREMAN:

WEATHER

F° _____ C° _____ _____ AM _____ PM

| HOURS DUE TO BAD WEATHER | ISSUED AND DELAYS |

NOTE: _______________________________________

| COMPLETION DATE | DAYS AHEAD OF SCHEDULE | DAYS BEHIND SCHEDULE |

SAFETY AND INCIDENTS

| SAFETY ISSUES THAT NEED TO BE ADDRESSED | ACCIDENTS / INCIDENTS / STEPS NEEDED TO RESOLVE |

SUMMARY OF THE WORK DONE TODAY

IMPORTANT NOTES

| NAME | SIGNATURE |

TODAY LABOR

INITIALS	TRADE	START	FINISH	PAID HOURS	OVERTIME	COMPANY
☐ EMPLOYEE ☐ CONTRUCTOR		AM	PM			
☐ EMPLOYEE ☐ CONTRUCTOR		AM	PM			
☐ EMPLOYEE ☐ CONTRUCTOR		AM	PM			
☐ EMPLOYEE ☐ CONTRUCTOR		AM	PM			
☐ EMPLOYEE ☐ CONTRUCTOR		AM	PM			
☐ EMPLOYEE ☐ CONTRUCTOR		AM	PM			
☐ EMPLOYEE ☐ CONTRUCTOR		AM	PM			
☐ EMPLOYEE ☐ CONTRUCTOR		AM	PM			

EQUIPMENT ON SITE	NO. OF UNITE	WORKING YES / NO

HIRED EQUIPMENT	NO. OF UNITE	EQUIPMENT RENTED	FROM	RATE

NAME: _______________________   SIGNATURE: _______________________

MO TU WE TH FR SA SU
☐   ☐   ☐   ☐   ☐   ☐   ☐

DATE: /   /

PROJECT:

FOREMAN:

WEATHER

F°____ C°____   ____ AM ____ PM

HOURS DUE TO
BAD WEATHER

ISSUED AND DELAYS

NOTE:

COMPLETION DATE	DAYS AHEAD OF SCHEDULE	DAYS BEHIND SCHEDULE

SAFETY AND INCIDENTS

SAFETY ISSUES THAT NEED TO BE ADDRESSED	ACCIDENTS / INCIDENTS / STEPS NEEDED TO RESOLVE

SUMMARY OF THE WORK DONE TODAY

IMPORTANT NOTES

NAME	SIGNATURE

TODAY LABOR

INITIALS	TRADE	START	FINISH	PAID HOURS	OVERTIME	COMPANY
☐ EMPLOYEE ☐ CONTRUCTOR		AM	PM			
☐ EMPLOYEE ☐ CONTRUCTOR		AM	PM			
☐ EMPLOYEE ☐ CONTRUCTOR		AM	PM			
☐ EMPLOYEE ☐ CONTRUCTOR		AM	PM			
☐ EMPLOYEE ☐ CONTRUCTOR		AM	PM			
☐ EMPLOYEE ☐ CONTRUCTOR		AM	PM			
☐ EMPLOYEE ☐ CONTRUCTOR		AM	PM			
☐ EMPLOYEE ☐ CONTRUCTOR		AM	PM			

EQUIPMENT ON SITE	NO. OF UNITE	WORKING YES / NO

HIRED EQUIPMENT	NO. OF UNITE	EQUIPMENT RENTED	FROM	RATE

NAME:

SIGNATURE:

MO TU WE TH FR SA SU
☐   ☐   ☐   ☐   ☐   ☐   ☐

DATE: /   /

PROJECT:

FOREMAN:

WEATHER

F°_____ C°_____   _____ AM _____ PM

HOURS DUE TO
BAD WEATHER

ISSUED AND DELAYS

NOTE: _______________________________

COMPLETION DATE	DAYS AHEAD OF SCHEDULE	DAYS BEHIND SCHEDULE

SAFETY AND INCIDENTS

SAFETY ISSUES THAT NEED TO BE ADDRESSED	ACCIDENTS / INCIDENTS / STEPS NEEDED TO RESOLVE

SUMMARY OF THE WORK DONE TODAY

IMPORTANT NOTES

NAME	SIGNATURE

TODAY LABOR

INITIALS	TRADE	START	FINISH	PAID HOURS	OVERTIME	COMPANY
☐ EMPLOYEE ☐ CONTRUCTOR		AM	PM			
☐ EMPLOYEE ☐ CONTRUCTOR		AM	PM			
☐ EMPLOYEE ☐ CONTRUCTOR		AM	PM			
☐ EMPLOYEE ☐ CONTRUCTOR		AM	PM			
☐ EMPLOYEE ☐ CONTRUCTOR		AM	PM			
☐ EMPLOYEE ☐ CONTRUCTOR		AM	PM			
☐ EMPLOYEE ☐ CONTRUCTOR		AM	PM			
☐ EMPLOYEE ☐ CONTRUCTOR		AM	PM			

EQUIPMENT ON SITE	NO. OF UNITE	WORKING YES / NO

HIRED EQUIPMENT	NO. OF UNITE	EQUIPMENT RENTED	FROM	RATE

NAME: _________________________   SIGNATURE: _________________________

MO   TU   WE   TH   FR   SA   SU
☐   ☐   ☐   ☐   ☐   ☐   ☐

DATE: /   /

PROJECT:

FOREMAN:

WEATHER

F°   C°   ___ AM   ___ PM

HOURS DUE TO BAD WEATHER	ISSUED AND DELAYS

NOTE:

COMPLETION DATE	DAYS AHEAD OF SCHEDULE	DAYS BEHIND SCHEDULE

SAFETY AND INCIDENTS

SAFETY ISSUES THAT NEED TO BE ADDRESSED	ACCIDENTS / INCIDENTS / STEPS NEEDED TO RESOLVE

SUMMARY OF THE WORK DONE TODAY

IMPORTANT NOTES

NAME	SIGNATURE

TODAY LABOR

INITIALS	TRADE	START	FINISH	PAID HOURS	OVERTIME	COMPANY
☐ EMPLOYEE ☐ CONTRUCTOR		AM	PM			
☐ EMPLOYEE ☐ CONTRUCTOR		AM	PM			
☐ EMPLOYEE ☐ CONTRUCTOR		AM	PM			
☐ EMPLOYEE ☐ CONTRUCTOR		AM	PM			
☐ EMPLOYEE ☐ CONTRUCTOR		AM	PM			
☐ EMPLOYEE ☐ CONTRUCTOR		AM	PM			
☐ EMPLOYEE ☐ CONTRUCTOR		AM	PM			
☐ EMPLOYEE ☐ CONTRUCTOR		AM	PM			

EQUIPMENT ON SITE	NO. OF UNITE	WORKING YES / NO

HIRED EQUIPMENT	NO. OF UNITE	EQUIPMENT RENTED	FROM	RATE

NAME: SIGNATURE:

MO   TU   WE   TH   FR   SA   SU
☐   ☐   ☐   ☐   ☐   ☐   ☐

DATE: /   /

PROJECT:

FOREMAN:

WEATHER

F°____   C°____   ____ AM   ____ PM

| HOURS DUE TO BAD WEATHER | ISSUED AND DELAYS |

NOTE:

COMPLETION DATE	DAYS AHEAD OF SCHEDULE	DAYS BEHIND SCHEDULE

SAFETY AND INCIDENTS

SAFETY ISSUES THAT NEED TO BE ADDRESSED	ACCIDENTS / INCIDENTS / STEPS NEEDED TO RESOLVE

SUMMARY OF THE WORK DONE TODAY

IMPORTANT NOTES

NAME	SIGNATURE

TODAY LABOR						
INITIALS	TRADE	START	FINISH	PAID HOURS	OVERTIME	COMPANY
☐ EMPLOYEE ☐ CONTRUCTOR		AM	PM			
☐ EMPLOYEE ☐ CONTRUCTOR		AM	PM			
☐ EMPLOYEE ☐ CONTRUCTOR		AM	PM			
☐ EMPLOYEE ☐ CONTRUCTOR		AM	PM			
☐ EMPLOYEE ☐ CONTRUCTOR		AM	PM			
☐ EMPLOYEE ☐ CONTRUCTOR		AM	PM			
☐ EMPLOYEE ☐ CONTRUCTOR		AM	PM			
☐ EMPLOYEE ☐ CONTRUCTOR		AM	PM			

EQUIPMENT ON SITE	NO. OF UNITE	WORKING YES / NO

HIRED EQUIPMENT	NO. OF UNITE	EQUIPMENT RENTED	FROM	RATE

NAME: ______________________ SIGNATURE: ______________________

MO TU WE TH FR SA SU
☐   ☐   ☐   ☐   ☐   ☐   ☐

DATE: ___ / ___ / ___

PROJECT:

FOREMAN:

WEATHER

F° ___   C° ___   ___ AM   ___ PM

HOURS DUE TO BAD WEATHER	ISSUED AND DELAYS

NOTE: _______________________

COMPLETION DATE	DAYS AHEAD OF SCHEDULE	DAYS BEHIND SCHEDULE

SAFETY AND INCIDENTS

SAFETY ISSUES THAT NEED TO BE ADDRESSED	ACCIDENTS / INCIDENTS / STEPS NEEDED TO RESOLVE

SUMMARY OF THE WORK DONE TODAY

IMPORTANT NOTES

NAME	SIGNATURE

TODAY LABOR

INITIALS	TRADE	START	FINISH	PAID HOURS	OVERTIME	COMPANY
☐ EMPLOYEE ☐ CONTRUCTOR		AM	PM			
☐ EMPLOYEE ☐ CONTRUCTOR		AM	PM			
☐ EMPLOYEE ☐ CONTRUCTOR		AM	PM			
☐ EMPLOYEE ☐ CONTRUCTOR		AM	PM			
☐ EMPLOYEE ☐ CONTRUCTOR		AM	PM			
☐ EMPLOYEE ☐ CONTRUCTOR		AM	PM			
☐ EMPLOYEE ☐ CONTRUCTOR		AM	PM			
☐ EMPLOYEE ☐ CONTRUCTOR		AM	PM			

EQUIPMENT ON SITE	NO. OF UNITE	WORKING YES / NO

HIRED EQUIPMENT	NO. OF UNITE	EQUIPMENT RENTED	FROM	RATE

NAME: SIGNATURE:

MO   TU   WE   TH   FR   SA   SU
☐   ☐   ☐   ☐   ☐   ☐   ☐

DATE: /   /

PROJECT: FOREMAN:

WEATHER   HOURS DUE TO BAD WEATHER   ISSUED AND DELAYS

F°   C°   AM   PM

NOTE:

COMPLETION DATE	DAYS AHEAD OF SCHEDULE	DAYS BEHIND SCHEDULE

SAFETY AND INCIDENTS

SAFETY ISSUES THAT NEED TO BE ADDRESSED	ACCIDENTS / INCIDENTS / STEPS NEEDED TO RESOLVE

SUMMARY OF THE WORK DONE TODAY

IMPORTANT NOTES

NAME	SIGNATURE

TODAY LABOR

INITIALS	TRADE	START	FINISH	PAID HOURS	OVERTIME	COMPANY
☐ EMPLOYEE ☐ CONTRUCTOR		AM	PM			
☐ EMPLOYEE ☐ CONTRUCTOR		AM	PM			
☐ EMPLOYEE ☐ CONTRUCTOR		AM	PM			
☐ EMPLOYEE ☐ CONTRUCTOR		AM	PM			
☐ EMPLOYEE ☐ CONTRUCTOR		AM	PM			
☐ EMPLOYEE ☐ CONTRUCTOR		AM	PM			
☐ EMPLOYEE ☐ CONTRUCTOR		AM	PM			
☐ EMPLOYEE ☐ CONTRUCTOR		AM	PM			

EQUIPMENT ON SITE	NO. OF UNITE	WORKING YES / NO

HIRED EQUIPMENT	NO. OF UNITE	EQUIPMENT RENTED	FROM	RATE

NAME: ___________________   SIGNATURE: ___________________

MO TU WE TH FR SA SU
☐   ☐   ☐   ☐   ☐   ☐   ☐

DATE: /   /

PROJECT:

FOREMAN:

WEATHER

F°   C°   AM   PM

HOURS DUE TO BAD WEATHER

ISSUED AND DELAYS

NOTE:

COMPLETION DATE	DAYS AHEAD OF SCHEDULE	DAYS BEHIND SCHEDULE

SAFETY AND INCIDENTS

SAFETY ISSUES THAT NEED TO BE ADDRESSED	ACCIDENTS / INCIDENTS / STEPS NEEDED TO RESOLVE

SUMMARY OF THE WORK DONE TODAY

IMPORTANT NOTES

NAME	SIGNATURE

TODAY LABOR

INITIALS	TRADE	START	FINISH	PAID HOURS	OVERTIME	COMPANY
☐ EMPLOYEE ☐ CONTRUCTOR		AM	PM			
☐ EMPLOYEE ☐ CONTRUCTOR		AM	PM			
☐ EMPLOYEE ☐ CONTRUCTOR		AM	PM			
☐ EMPLOYEE ☐ CONTRUCTOR		AM	PM			
☐ EMPLOYEE ☐ CONTRUCTOR		AM	PM			
☐ EMPLOYEE ☐ CONTRUCTOR		AM	PM			
☐ EMPLOYEE ☐ CONTRUCTOR		AM	PM			
☐ EMPLOYEE ☐ CONTRUCTOR		AM	PM			

EQUIPMENT ON SITE	NO. OF UNITE	WORKING YES / NO

HIRED EQUIPMENT	NO. OF UNITE	EQUIPMENT RENTED	FROM	RATE

NAME: _______________________   SIGNATURE: _______________________

MO TU WE TH FR SA SU

DATE: /   /

PROJECT:

FOREMAN:

WEATHER

F°   C°   AM   PM

HOURS DUE TO BAD WEATHER

ISSUED AND DELAYS

NOTE:

COMPLETION DATE	DAYS AHEAD OF SCHEDULE	DAYS BEHIND SCHEDULE

SAFETY AND INCIDENTS

SAFETY ISSUES THAT NEED TO BE ADDRESSED	ACCIDENTS / INCIDENTS / STEPS NEEDED TO RESOLVE

SUMMARY OF THE WORK DONE TODAY

IMPORTANT NOTES

NAME	SIGNATURE

TODAY LABOR

INITIALS	TRADE	START	FINISH	PAID HOURS	OVERTIME	COMPANY
☐ EMPLOYEE ☐ CONTRUCTOR		AM	PM			
☐ EMPLOYEE ☐ CONTRUCTOR		AM	PM			
☐ EMPLOYEE ☐ CONTRUCTOR		AM	PM			
☐ EMPLOYEE ☐ CONTRUCTOR		AM	PM			
☐ EMPLOYEE ☐ CONTRUCTOR		AM	PM			
☐ EMPLOYEE ☐ CONTRUCTOR		AM	PM			
☐ EMPLOYEE ☐ CONTRUCTOR		AM	PM			
☐ EMPLOYEE ☐ CONTRUCTOR		AM	PM			

EQUIPMENT ON SITE	NO. OF UNITE	WORKING YES / NO

HIRED EQUIPMENT	NO. OF UNITE	EQUIPMENT RENTED	FROM	RATE

NAME: ______________________   SIGNATURE: ______________________

MO   TU   WE   TH   FR   SA   SU
☐   ☐   ☐   ☐   ☐   ☐   ☐

DATE: /   /

PROJECT:

FOREMAN:

WEATHER

F° _____   C° _____   _____ AM   _____ PM

HOURS DUE TO BAD WEATHER	ISSUED AND DELAYS

NOTE: ____

COMPLETION DATE	DAYS AHEAD OF SCHEDULE	DAYS BEHIND SCHEDULE

SAFETY AND INCIDENTS

SAFETY ISSUES THAT NEED TO BE ADDRESSED	ACCIDENTS / INCIDENTS / STEPS NEEDED TO RESOLVE

SUMMARY OF THE WORK DONE TODAY

IMPORTANT NOTES

NAME	SIGNATURE

TODAY LABOR						
INITIALS	TRADE	START	FINISH	PAID HOURS	OVERTIME	COMPANY
☐ EMPLOYEE ☐ CONTRUCTOR		AM	PM			
☐ EMPLOYEE ☐ CONTRUCTOR		AM	PM			
☐ EMPLOYEE ☐ CONTRUCTOR		AM	PM			
☐ EMPLOYEE ☐ CONTRUCTOR		AM	PM			
☐ EMPLOYEE ☐ CONTRUCTOR		AM	PM			
☐ EMPLOYEE ☐ CONTRUCTOR		AM	PM			
☐ EMPLOYEE ☐ CONTRUCTOR		AM	PM			
☐ EMPLOYEE ☐ CONTRUCTOR		AM	PM			

EQUIPMENT ON SITE	NO. OF UNITE	WORKING YES / NO

HIRED EQUIPMENT	NO. OF UNITE	EQUIPMENT RENTED	FROM	RATE

NAME: ______________________   SIGNATURE: ______________________

MO TU WE TH FR SA SU
☐   ☐   ☐   ☐   ☐   ☐   ☐

DATE: ___ / ___ / ___

PROJECT: _______________________

FOREMAN: _______________________

WEATHER

F° ______   C° ______   ______ AM   ______ PM

HOURS DUE TO BAD WEATHER	ISSUED AND DELAYS

NOTE: _______________________

COMPLETION DATE	DAYS AHEAD OF SCHEDULE	DAYS BEHIND SCHEDULE

SAFETY AND INCIDENTS

SAFETY ISSUES THAT NEED TO BE ADDRESSED	ACCIDENTS / INCIDENTS / STEPS NEEDED TO RESOLVE

SUMMARY OF THE WORK DONE TODAY

IMPORTANT NOTES

NAME	SIGNATURE

TODAY LABOR

INITIALS	TRADE	START	FINISH	PAID HOURS	OVERTIME	COMPANY
☐ EMPLOYEE ☐ CONTRUCTOR		AM	PM			
☐ EMPLOYEE ☐ CONTRUCTOR		AM	PM			
☐ EMPLOYEE ☐ CONTRUCTOR		AM	PM			
☐ EMPLOYEE ☐ CONTRUCTOR		AM	PM			
☐ EMPLOYEE ☐ CONTRUCTOR		AM	PM			
☐ EMPLOYEE ☐ CONTRUCTOR		AM	PM			
☐ EMPLOYEE ☐ CONTRUCTOR		AM	PM			
☐ EMPLOYEE ☐ CONTRUCTOR		AM	PM			

EQUIPMENT ON SITE	NO. OF UNITE	WORKING YES / NO

HIRED EQUIPMENT	NO. OF UNITE	EQUIPMENT RENTED	FROM	RATE

NAME: _______________________   SIGNATURE: _______________________

MO	TU	WE	TH	FR	SA	SU		DATE: / /
☐	☐	☐	☐	☐	☐	☐		

PROJECT:

FOREMAN:

WEATHER F°___ C°___ ___ AM ___ PM

HOURS DUE TO BAD WEATHER	ISSUED AND DELAYS

NOTE:

COMPLETION DATE	DAYS AHEAD OF SCHEDULE	DAYS BEHIND SCHEDULE

SAFETY AND INCIDENTS

SAFETY ISSUES THAT NEED TO BE ADDRESSED	ACCIDENTS / INCIDENTS / STEPS NEEDED TO RESOLVE

SUMMARY OF THE WORK DONE TODAY

IMPORTANT NOTES

NAME	SIGNATURE

TODAY LABOR

INITIALS	TRADE	START	FINISH	PAID HOURS	OVERTIME	COMPANY
☐ EMPLOYEE ☐ CONTRUCTOR		AM	PM			
☐ EMPLOYEE ☐ CONTRUCTOR		AM	PM			
☐ EMPLOYEE ☐ CONTRUCTOR		AM	PM			
☐ EMPLOYEE ☐ CONTRUCTOR		AM	PM			
☐ EMPLOYEE ☐ CONTRUCTOR		AM	PM			
☐ EMPLOYEE ☐ CONTRUCTOR		AM	PM			
☐ EMPLOYEE ☐ CONTRUCTOR		AM	PM			
☐ EMPLOYEE ☐ CONTRUCTOR		AM	PM			

EQUIPMENT ON SITE	NO. OF UNITE	WORKING YES / NO

HIRED EQUIPMENT	NO. OF UNITE	EQUIPMENT RENTED	FROM	RATE

NAME: ___________________   SIGNATURE: ___________________

MO   TU   WE   TH   FR   SA   SU
☐   ☐   ☐   ☐   ☐   ☐   ☐

DATE: /   /

PROJECT:

FOREMAN:

WEATHER

F° _____   C° _____   _____ AM   _____ PM

HOURS DUE TO BAD WEATHER	ISSUED AND DELAYS

NOTE: ____

COMPLETION DATE	DAYS AHEAD OF SCHEDULE	DAYS BEHIND SCHEDULE

SAFETY AND INCIDENTS

SAFETY ISSUES THAT NEED TO BE ADDRESSED	ACCIDENTS / INCIDENTS / STEPS NEEDED TO RESOLVE

SUMMARY OF THE WORK DONE TODAY

IMPORTANT NOTES

NAME	SIGNATURE

TODAY LABOR

INITIALS	TRADE	START	FINISH	PAID HOURS	OVERTIME	COMPANY
☐ EMPLOYEE ☐ CONTRUCTOR		AM	PM			
☐ EMPLOYEE ☐ CONTRUCTOR		AM	PM			
☐ EMPLOYEE ☐ CONTRUCTOR		AM	PM			
☐ EMPLOYEE ☐ CONTRUCTOR		AM	PM			
☐ EMPLOYEE ☐ CONTRUCTOR		AM	PM			
☐ EMPLOYEE ☐ CONTRUCTOR		AM	PM			
☐ EMPLOYEE ☐ CONTRUCTOR		AM	PM			
☐ EMPLOYEE ☐ CONTRUCTOR		AM	PM			

EQUIPMENT ON SITE	NO. OF UNITE	WORKING YES / NO

HIRED EQUIPMENT	NO. OF UNITE	EQUIPMENT RENTED	FROM	RATE

NAME: ___________________________ SIGNATURE: ___________________________

MO   TU   WE   TH   FR   SA   SU
☐   ☐   ☐   ☐   ☐   ☐   ☐

DATE: /   /

PROJECT:

FOREMAN:

WEATHER

F°______   C°______   ______ AM   ______ PM

| HOURS DUE TO BAD WEATHER | ISSUED AND DELAYS |

NOTE:

COMPLETION DATE	DAYS AHEAD OF SCHEDULE	DAYS BEHIND SCHEDULE

SAFETY AND INCIDENTS

SAFETY ISSUES THAT NEED TO BE ADDRESSED	ACCIDENTS / INCIDENTS / STEPS NEEDED TO RESOLVE

SUMMARY OF THE WORK DONE TODAY

IMPORTANT NOTES

NAME	SIGNATURE

TODAY LABOR

INITIALS	TRADE	START	FINISH	PAID HOURS	OVERTIME	COMPANY
☐ EMPLOYEE ☐ CONTRUCTOR		AM	PM			
☐ EMPLOYEE ☐ CONTRUCTOR		AM	PM			
☐ EMPLOYEE ☐ CONTRUCTOR		AM	PM			
☐ EMPLOYEE ☐ CONTRUCTOR		AM	PM			
☐ EMPLOYEE ☐ CONTRUCTOR		AM	PM			
☐ EMPLOYEE ☐ CONTRUCTOR		AM	PM			
☐ EMPLOYEE ☐ CONTRUCTOR		AM	PM			
☐ EMPLOYEE ☐ CONTRUCTOR		AM	PM			

EQUIPMENT ON SITE	NO. OF UNITE	WORKING YES / NO

HIRED EQUIPMENT	NO. OF UNITE	EQUIPMENT RENTED	FROM	RATE

NAME: SIGNATURE:

MO TU WE TH FR SA SU

DATE: / /

PROJECT:

FOREMAN:

WEATHER

F°   C°   AM   PM

HOURS DUE TO
BAD WEATHER

ISSUED AND DELAYS

NOTE:

COMPLETION DATE	DAYS AHEAD OF SCHEDULE	DAYS BEHIND SCHEDULE

SAFETY AND INCIDENTS

SAFETY ISSUES THAT NEED TO BE ADDRESSED	ACCIDENTS / INCIDENTS / STEPS NEEDED TO RESOLVE

SUMMARY OF THE WORK DONE TODAY

IMPORTANT NOTES

NAME	SIGNATURE

TODAY LABOR

INITIALS	TRADE	START	FINISH	PAID HOURS	OVERTIME	COMPANY
☐ EMPLOYEE ☐ CONTRUCTOR		AM	PM			
☐ EMPLOYEE ☐ CONTRUCTOR		AM	PM			
☐ EMPLOYEE ☐ CONTRUCTOR		AM	PM			
☐ EMPLOYEE ☐ CONTRUCTOR		AM	PM			
☐ EMPLOYEE ☐ CONTRUCTOR		AM	PM			
☐ EMPLOYEE ☐ CONTRUCTOR		AM	PM			
☐ EMPLOYEE ☐ CONTRUCTOR		AM	PM			
☐ EMPLOYEE ☐ CONTRUCTOR		AM	PM			

EQUIPMENT ON SITE	NO. OF UNITE	WORKING YES / NO

HIRED EQUIPMENT	NO. OF UNITE	EQUIPMENT RENTED	FROM	RATE

NAME: _____________________   SIGNATURE: _____________________

MO   TU   WE   TH   FR   SA   SU
☐   ☐   ☐   ☐   ☐   ☐   ☐

DATE: /   /

PROJECT:

FOREMAN:

WEATHER

F°____ C°____   ____ AM   ____ PM

HOURS DUE TO BAD WEATHER	ISSUED AND DELAYS

NOTE: ______________________________________

COMPLETION DATE	DAYS AHEAD OF SCHEDULE	DAYS BEHIND SCHEDULE

SAFETY AND INCIDENTS

SAFETY ISSUES THAT NEED TO BE ADDRESSED	ACCIDENTS / INCIDENTS / STEPS NEEDED TO RESOLVE

SUMMARY OF THE WORK DONE TODAY

IMPORTANT NOTES

NAME	SIGNATURE

TODAY LABOR

INITIALS	TRADE	START	FINISH	PAID HOURS	OVERTIME	COMPANY
☐ EMPLOYEE ☐ CONTRUCTOR		AM	PM			
☐ EMPLOYEE ☐ CONTRUCTOR		AM	PM			
☐ EMPLOYEE ☐ CONTRUCTOR		AM	PM			
☐ EMPLOYEE ☐ CONTRUCTOR		AM	PM			
☐ EMPLOYEE ☐ CONTRUCTOR		AM	PM			
☐ EMPLOYEE ☐ CONTRUCTOR		AM	PM			
☐ EMPLOYEE ☐ CONTRUCTOR		AM	PM			
☐ EMPLOYEE ☐ CONTRUCTOR		AM	PM			

EQUIPMENT ON SITE	NO. OF UNITE	WORKING YES / NO

HIRED EQUIPMENT	NO. OF UNITE	EQUIPMENT RENTED	FROM	RATE

NAME: _______________________   SIGNATURE: _______________________

MO TU WE TH FR SA SU

DATE: / /

PROJECT:

FOREMAN:

WEATHER

F° C° _____ AM _____ PM

HOURS DUE TO BAD WEATHER

ISSUED AND DELAYS

NOTE: _____

COMPLETION DATE	DAYS AHEAD OF SCHEDULE	DAYS BEHIND SCHEDULE

SAFETY AND INCIDENTS

SAFETY ISSUES THAT NEED TO BE ADDRESSED	ACCIDENTS / INCIDENTS / STEPS NEEDED TO RESOLVE

SUMMARY OF THE WORK DONE TODAY

IMPORTANT NOTES

NAME	SIGNATURE

TODAY LABOR

INITIALS	TRADE	START	FINISH	PAID HOURS	OVERTIME	COMPANY
☐ EMPLOYEE ☐ CONTRUCTOR		AM	PM			
☐ EMPLOYEE ☐ CONTRUCTOR		AM	PM			
☐ EMPLOYEE ☐ CONTRUCTOR		AM	PM			
☐ EMPLOYEE ☐ CONTRUCTOR		AM	PM			
☐ EMPLOYEE ☐ CONTRUCTOR		AM	PM			
☐ EMPLOYEE ☐ CONTRUCTOR		AM	PM			
☐ EMPLOYEE ☐ CONTRUCTOR		AM	PM			
☐ EMPLOYEE ☐ CONTRUCTOR		AM	PM			

EQUIPMENT ON SITE	NO. OF UNITE	WORKING YES / NO

HIRED EQUIPMENT	NO. OF UNITE	EQUIPMENT RENTED	FROM	RATE

NAME: _______________________   SIGNATURE: _______________________

MO TU WE TH FR SA SU
☐   ☐   ☐   ☐   ☐   ☐   ☐

DATE: ___ / ___ / ___

PROJECT:

FOREMAN:

WEATHER

F° _____ C° _____   _____ AM _____ PM

HOURS DUE TO
BAD WEATHER

ISSUED AND DELAYS

NOTE: ______________________________

COMPLETION DATE	DAYS AHEAD OF SCHEDULE	DAYS BEHIND SCHEDULE

SAFETY AND INCIDENTS

SAFETY ISSUES THAT NEED TO BE ADDRESSED	ACCIDENTS / INCIDENTS / STEPS NEEDED TO RESOLVE

SUMMARY OF THE WORK DONE TODAY

IMPORTANT NOTES

NAME	SIGNATURE

TODAY LABOR

INITIALS	TRADE	START	FINISH	PAID HOURS	OVERTIME	COMPANY
☐ EMPLOYEE ☐ CONTRUCTOR		AM	PM			
☐ EMPLOYEE ☐ CONTRUCTOR		AM	PM			
☐ EMPLOYEE ☐ CONTRUCTOR		AM	PM			
☐ EMPLOYEE ☐ CONTRUCTOR		AM	PM			
☐ EMPLOYEE ☐ CONTRUCTOR		AM	PM			
☐ EMPLOYEE ☐ CONTRUCTOR		AM	PM			
☐ EMPLOYEE ☐ CONTRUCTOR		AM	PM			
☐ EMPLOYEE ☐ CONTRUCTOR		AM	PM			

EQUIPMENT ON SITE	NO. OF UNITE	WORKING YES / NO

HIRED EQUIPMENT	NO. OF UNITE	EQUIPMENT RENTED	FROM	RATE

NAME: ______________________   SIGNATURE: ______________________

MO   TU   WE   TH   FR   SA   SU   ☐   ☐   ☐   ☐   ☐   ☐   ☐

DATE: /   /

PROJECT:

FOREMAN:

WEATHER

F°   C°   AM   PM

HOURS DUE TO BAD WEATHER

ISSUED AND DELAYS

NOTE:

COMPLETION DATE	DAYS AHEAD OF SCHEDULE	DAYS BEHIND SCHEDULE

SAFETY AND INCIDENTS

SAFETY ISSUES THAT NEED TO BE ADDRESSED	ACCIDENTS / INCIDENTS / STEPS NEEDED TO RESOLVE

SUMMARY OF THE WORK DONE TODAY

IMPORTANT NOTES

NAME	SIGNATURE

TODAY LABOR

INITIALS	TRADE	START	FINISH	PAID HOURS	OVERTIME	COMPANY
☐ EMPLOYEE ☐ CONTRUCTOR		AM	PM			
☐ EMPLOYEE ☐ CONTRUCTOR		AM	PM			
☐ EMPLOYEE ☐ CONTRUCTOR		AM	PM			
☐ EMPLOYEE ☐ CONTRUCTOR		AM	PM			
☐ EMPLOYEE ☐ CONTRUCTOR		AM	PM			
☐ EMPLOYEE ☐ CONTRUCTOR		AM	PM			
☐ EMPLOYEE ☐ CONTRUCTOR		AM	PM			
☐ EMPLOYEE ☐ CONTRUCTOR		AM	PM			

EQUIPMENT ON SITE	NO. OF UNITE	WORKING YES / NO

HIRED EQUIPMENT	NO. OF UNITE	EQUIPMENT RENTED	FROM	RATE

NAME: _______________________   SIGNATURE: _______________________

MO   TU   WE   TH   FR   SA   SU
☐   ☐   ☐   ☐   ☐   ☐   ☐

DATE: ___ / ___ / ___

PROJECT:

FOREMAN:

WEATHER   F° _____   C° _____   _____ AM   _____ PM

HOURS DUE TO BAD WEATHER	ISSUED AND DELAYS

NOTE: ____

COMPLETION DATE	DAYS AHEAD OF SCHEDULE	DAYS BEHIND SCHEDULE

SAFETY AND INCIDENTS

SAFETY ISSUES THAT NEED TO BE ADDRESSED	ACCIDENTS / INCIDENTS / STEPS NEEDED TO RESOLVE

SUMMARY OF THE WORK DONE TODAY

IMPORTANT NOTES

NAME	SIGNATURE

<table>
<tr><td colspan="8" align="center">TODAY LABOR</td></tr>
</table>

INITIALS	TRADE	START	FINISH	PAID HOURS	OVERTIME	COMPANY
☐ EMPLOYEE ☐ CONTRUCTOR		AM	PM			
☐ EMPLOYEE ☐ CONTRUCTOR		AM	PM			
☐ EMPLOYEE ☐ CONTRUCTOR		AM	PM			
☐ EMPLOYEE ☐ CONTRUCTOR		AM	PM			
☐ EMPLOYEE ☐ CONTRUCTOR		AM	PM			
☐ EMPLOYEE ☐ CONTRUCTOR		AM	PM			
☐ EMPLOYEE ☐ CONTRUCTOR		AM	PM			
☐ EMPLOYEE ☐ CONTRUCTOR		AM	PM			

EQUIPMENT ON SITE	NO. OF UNITE	WORKING YES / NO

HIRED EQUIPMENT	NO. OF UNITE	EQUIPMENT RENTED	FROM	RATE

NAME: _______________________   SIGNATURE: _______________________

MO   TU   WE   TH   FR   SA   SU
☐   ☐   ☐   ☐   ☐   ☐   ☐

DATE: /   /

PROJECT:

FOREMAN:

WEATHER

F°_______ C°_______   _______ AM   _______ PM

| HOURS DUE TO BAD WEATHER | ISSUED AND DELAYS |

NOTE:

| COMPLETION DATE | DAYS AHEAD OF SCHEDULE | DAYS BEHIND SCHEDULE |

SAFETY AND INCIDENTS

| SAFETY ISSUES THAT NEED TO BE ADDRESSED | ACCIDENTS / INCIDENTS / STEPS NEEDED TO RESOLVE |

SUMMARY OF THE WORK DONE TODAY

IMPORTANT NOTES

| NAME | SIGNATURE |

TODAY LABOR

INITIALS	TRADE	START	FINISH	PAID HOURS	OVERTIME	COMPANY
☐ EMPLOYEE ☐ CONTRUCTOR		AM	PM			
☐ EMPLOYEE ☐ CONTRUCTOR		AM	PM			
☐ EMPLOYEE ☐ CONTRUCTOR		AM	PM			
☐ EMPLOYEE ☐ CONTRUCTOR		AM	PM			
☐ EMPLOYEE ☐ CONTRUCTOR		AM	PM			
☐ EMPLOYEE ☐ CONTRUCTOR		AM	PM			
☐ EMPLOYEE ☐ CONTRUCTOR		AM	PM			
☐ EMPLOYEE ☐ CONTRUCTOR		AM	PM			

EQUIPMENT ON SITE	NO. OF UNITE	WORKING YES / NO

HIRED EQUIPMENT	NO. OF UNITE	EQUIPMENT RENTED	FROM	RATE

NAME: ___________________ SIGNATURE: ___________________

IMPORTANT TELEPHONE NUMBER

▶ NAME _______________________________ PHONE _______________________
  EMAIL ____

▶ NAME _______________________________ PHONE _______________________
  EMAIL ____

▶ NAME _______________________________ PHONE _______________________
  EMAIL ____

▶ NAME _______________________________ PHONE _______________________
  EMAIL ____

▶ NAME _______________________________ PHONE _______________________
  EMAIL ____

▶ NAME _______________________________ PHONE _______________________
  EMAIL ____

▶ NAME _______________________________ PHONE _______________________
  EMAIL ____

▶ NAME _______________________________ PHONE _______________________
  EMAIL ____

▶ NAME _______________________________ PHONE _______________________
  EMAIL ____

▶ NAME _______________________________ PHONE _______________________
  EMAIL ____

▶ NAME _______________________________ PHONE _______________________
  EMAIL ____

▶ NAME _______________________________ PHONE _______________________
  EMAIL ____

▶ NAME _______________________________ PHONE _______________________
  EMAIL ____